The Buncefield Explosion

A Compilation by
Sceptre Fundraising Team

PUBLISHED BY SCEPTRE EDUCATION

COPYRIGHT

© SCEPTRE EDUCATION 2006

10062999

Produced in Great Britain for Sceptre Education.

ISBN-10 0-9552759-0-3 (Valid until 1/1/07)
ISBN-13 978-0-9552759-0-6

SCEPTRE SCHOOL
RIDGEWAY AVENUE, DUNSTABLE, BEDS LU5 4QL

Acknowledgements

The following have kindly donated material for use within this book.

Jon Smith	– Station Commander, Hemel Fire Station
Jeremy Harris	– Firefighter, St Albans Fire Station
Jon Batchelor	– Blue Watch Commander, Hemel Fire Station
Blue Watch	– Hemel Fire Station
Gordon MacMillan	– Buncefield Review Team
Joe Cartwright	– Hemel Ambulance Station
Claire Ling	– Hertfordshire Police
HSE	– Buncefield Major Incident Investigation Board
John O'Reilley	– South Beds News
Ian Silverstein	– Local Resident
Craig Shepheard	– Shepheard Photography
David Fieldstein	– Editor, Hemel Hempstead Gazette
Rebecca Barnard	– Associated Newspapers
Martin Keen	– Press Association
Dawn O'Driscoll	– © The Daily Telegraph 2005 – John Taylor
Kim Lee	– Reuters Media
Kevin Collins	– The Express Newspaper Group
Ben Cawthra	– INS Newsgroup Ltd
Captain Michael Bainbridge	– Salvation Army
Julian Coyne	– BPA
Nigel Marsh	– Fuji
Zachary Phole	– Furnell Transport
Richard Stewart	– Photographer
Leigh Quinnell	– Photographer
Nick Ray	– Photographer
Eddie Orija	– Science Teacher, Sceptre School
SCOOPT.COM	
Chiltern Air Support Unit	

The Sceptre School fundraising team wishes to thank everyone who has kindly donated material for use within this book. Every effort has been made to trace copyright holders but, in a few cases, this has proved impossible. The publishers therefore wish to thank the copyright holders of these photos and to assure them that the use of their material has benefitted registered charities.

INTRODUCTION

by
Sub Officer Jon Batchelor,
Blue Watch Commander,
Hemel Hempstead Fire Station
First firefighter on the scene

SATURDAY 10th of December 2005 was unusual only in the fact that Hemel Hempstead Fire Station had received no fire calls during the night. I awoke at 05:45 Sunday, and was peacefully dozing, when at 06:01 the explosion literally shook the station, its sound resonating all around then fading into the distance. My initial thought was that a thunder clap had occurred directly above the station, but this soon changed as I heard footsteps running down the stairs followed by the sounding of the local alarm system. (Ff Everett had seen the fire-ball going up from the mess deck window.) We were getting dressed into fire gear as Control turned us out to 'Explosion Maylands Avenue, rear of Masons Road.'

We proceeded up Queensway and the view made me think of a scene which my father had described to me when I was a small boy, of how the horizon glowed red when London was being bombed during the blitz, but this was Sunday 11th December 2005, and London was in the opposite direction!

As we entered the industrial area we could see flames climbing hundreds of feet into the dark sky producing an eerie false daylight, but we were still unsure of the exact location of the fire. As we entered Boundary Way we were confronted by a scene of utter devastation, with buildings severely damaged and glass and debris lying across the road. It was passing the badly-damaged Fuji building that I was able to look through a gap and got my first glimpse of the many fuel tanks silhouetted against the raging fire.

It was Buncefield. It was the big one!! It was not just part of Buncefield – it appeared to be the whole of Buncefield and it was totally engulfed in flames!

I informed Control that the incident was Buncefield Oil Terminal and declared it a major incident. To declare a major incident is probably the biggest call which a Fire Officer can ever make, especially one of my lowly rank, but the decision was such an easy one to take. It was simply the biggest fire and scene of complete devastation that I had ever seen! I feel confident that any other officer in my position would have come to the same conclusion.

The events which unfolded during those first few hours, and the subsequent efforts of all of those who came together to fight and extinguish the largest fire in Europe for 60 years, will be described in the pages of this book in infinitely more eloquent and articulate words than I can aspire to and I thank the authors for their time and effort in telling our story.

I would like to take this opportunity to express my deep pride and admiration for the members of Blue Watch, Hemel Hempstead for their unfaltering professionalism, tenacity and bravery during the opening moments of the incident and also all our colleagues both in Hertfordshire and from Brigades around the country, who through their blood, sweat and toil, managed to bring the incident to a successful conclusion.

This, of course, could not have been possible without the superb efforts of the senior officers who manned the Gold, Silver and Bronze command posts and who planned, calculated, organized and implemented one of the biggest logistical operations ever undertaken by the British Fire Service.

Buncefield was for me a life-changing experience and I will remember it for the rest of my life. The enduring image of the thick black velvet smoke rolling up into the beautiful blue winter's morning sky and spreading out like a huge apocalyptic shroud will be etched onto the memory of all who viewed it. It has brought a very close group of men closer still and I will finish by saying "We were there. We got the big one!!!!"

Hemel Hempstead, Hertfordshire:
25 miles North West of Central London: Population 84,000.

Since July 1968 'Hemel' had hosted a large oil storage facility on the outskirts of the town. Despite over 400 road tanker collections daily most residents thought little of the depot's presence and some were completely unaware of it.

Very few saw it as a threat.

Another normal day, another delivery to a local filling station after an early morning collection from Buncefield Oil Storage and Distribution Depot.

The town's famous "magic roundabout" and other streets are almost deserted....

05:40

Local residents sleep peacefully, unaware that only a short distance across the fields a virtual time bomb is ticking....

Blue Watch, headed by Jon Batchelor, are on duty. So far there has not been a single call during the night...

However a national disaster is imminent...

At Buncefield Depot a sinister vapour cloud appeared in the site floodlights. By 06:00 this thick mist had intensified behind storage tanks and rolled forwards to the loading gantry and backwards across the boundary fences into adjacent office car parks. Witnesses noticed a very strong smell of fuel, and a nearby motorist gave reports of his engine racing, then stalling after he drove into a 'murky cloud'.

Tanker drivers said: "We switched off tanker lights and made for the exit gates..."

06:01

Now spread over several acres, the vapour cloud ignited, triggering a massive aerial explosion preceded by violent ground shuddering. As shock waves ripped through the whole town and beyond, enormous fire-balls were flung into the air, engulfing buildings and fuel tanks. Damaged tanks caught fire instantly in a relentless chain reaction.

Captured

The exact moment of the blast is captured on two

11/12/2005 06:04:41
Lobby - Furnell

11/12/2005 06:04:46
Lobby - Furnell

Inside cameras show doors and windows being blown inwards by the blast.

on Camera

seconds of CCTV footage at a nearby transport company.

Outside cameras show the early-morning darkness transformed into near-daylight.

Thousands were shaken awake in the darkness by the loudest explosion most would ever experience. Windows and doors were smashed or wrenched off, roofs were shattered and fixtures and contents of rooms were hurled about. Nearby residents were thrown from their beds or covered by debris where they lay.

Hundreds fled from their houses and gathered on the streets in shock and disbelief, convinced that there had been a terrorist attack or a plane crash. There was a wild medley of car alarms, house alarms, barking dogs and frightened people, while on the horizon the pitch darkness was broken by the soaring fire-balls and further explosions.

At Golden West Foods, 1/4 mile from the Depot, a security guard in the gatehouse took a phone call from a delivery contractor at 06:00. "At exactly 6 o'clock he was giving me his details, so I was sitting writing with my head down... then that massive blast sent the window completely over my head... the glass was stuck on the opposite wall like a knife. That contractor saved my life – I would have been cut in half..."

"It was sirens, sirens, as if every emergency vehicle in the town was on the road within minutes…"

STATION COPY
Message from Standby Modes &19.3.1@
06:04:51 on 11 DEC 05

-
                        TURNOUT
                         011
-

  Turnout 011

  Address MAYLANDS AVENUE,
           REAR OF MASONS ROAD

     Type GAS EXPLOSION

  Map Ref 505811,206906

   Caller 01727200004

  Details EXPLOSION-MAYLANDS AVENUE,
           REAR OF MASONS ROAD

Map Book Page 124 D2

Incident 46469

--*-*-*-*-*-*-*   END   *-*-*-*-*-*-*-*-*-*-*

"It was absolute panic and terror in t
– people milling around, some half-dr
some in stunned silence, some in a da
walking anywhere, even towards that
horizon. Was it terrorism? Was it nucl
No-one knew."

"It was more like a weekday rush-hour than 6:00 am on a Sunday… people roared off in their cars – they just had to go and see what had happened…"

Within seconds of the explosion local emergency services were in action. Firefighters raced out to the north-east of the town, taking more specific directions en-route. Emergency calls poured in, with initial confusion as to the exact location of the incident. Buncefield Depot was quickly pinpointed – "The Big One."

Police teams moved in quickly to establish cordons, evacuate over 2000 people from houses and business premises and assist with the logistics of the Fire and Rescue operation.

"We were told to 'get out, just get out'… and driving away from that horror I met people heading towards the danger I was escaping from!"
(Factory night shift worker)

"Like nothing they had

"*The sky was glowing red on the northern horizon, something was wrong – too early for sunrise, wrong direction anyway, it was glowing brighter very fast...*"

"*Suddenly things were flying around... things coming towards me... I remember a blinding flash...*" (BP Service Station, duty manager)

"*A huge flame gushed up into the sky, what on earth...? raced through my mind: I could hardly believe that what I was seeing was real.*"

"*I thought lightning had struck the balcony at first... then I realised it was a bit further away... a plane crash!*"

"*...We could hear car alarms sounding, people were pointing northwards, the sky was orangey-red and flames were leaping above the trees, about 150 foot high, it was terrifying to watch.*"

ever seen or heard before."

"...I thought it was an earthquake... the house badly shook before the explosion, it was the shake that woke us up... you're half awake and half asleep, don't know what it's about and then there's this big bang..."

"...a rumble swelling to a colossal roar, louder than the loudest thunderclap..."

"...I thought it was a nuclear explosion... I was expecting more of a blast to come after the boom."

"...I wanted to pray because I was scared, but Dad and Mum's bed had dust on it and I had to look for a place that wasn't dusty..." (local six year old)

"...It was like the noise that planes make when they break the sound barrier out at sea... or when an old WWII mine is found and they blow it up." (Phil Smith, Aldeburgh, Suffolk Coast)

21

INFERNO

This is the scene
that greeted the
first firefighters on
arrival. There are
over 100 million
litres of fuel on
fire. Tanks are
exploding and the
fire is spreading
fast. All dedicated
on-site fire
fighting
equipment and
most of the water
supplies have been
destroyed by the
explosion.

And 6 men are missing...

FIRST PRIORITY

"Once I realised it was Buncefield I made it a major incident – it was obviously far in excess of anything we could handle ourselves as a service..."

"I was trying to get information but it wasn't possible – this bloke stood there in utter shock, like he was concussed, saying there had been 6 men in there, but we didn't know where..."

"First of all we went up as far as the loading gantries – it was as near as you could get to the fire because of the heat. I was convinced we'd find persons in the tanker cabs or laid out on the tarmac, but we didn't..."

"There were a few people staggering around with cuts and bruises – they were rushed to hospital in cars by members of the public."

"We couldn't get near the fire, we didn't have the resources to even start to fight it. Our priority was looking for life – injured life, lost life..."

"Then we checked the site offices... just had to smash your way into each because the doors were jammed by debris... it was eerie, you would find a wrecked desk and chair, a fluorescent jacket on the chairback so the chap could be in there under it all... but when you searched – nobody! And every office we went into we expected to find people and never did..."

SEARCH FOR LIFE

As the search for life was proceeding, crews were alerted to a secondary fire on all three storeys at the back of Northgate IT. Although small in comparison with the adjacent depot inferno this was in fact a major threat to the whole building and the surrounding industrial complex. Eight fire engines and crews from Redbourn and St Albans were assigned to this blaze and it took most of Sunday to extinguish it.

Nothing could have prepared local fire-fighters for this. Buncefield had been equipped and trained for fire, even a major fire, but the explosion had wrecked most of the resources. Huge quantities of foam concentrate were needed; many more fire crews were needed; a new water source was needed.

The national major incident command response was already being mobilised, but little could be achieved until it was all in place.

Meanwhile, the fire raged.

Hemel Fire Station

Piccotts End

Grovehill

Highfield

Magic Roundabout

Hemel Hospital

34

Nearest Residents

Buncefield Depot

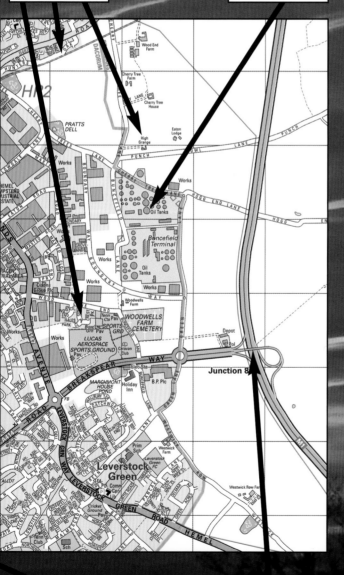

Hemel Hempstead bravely faced up to a period of chaos and uncertainty as the town was catapulted into the national spotlight. Major roads were closed indefinitely, shops were deserted, schools were shut and thousands faced the prospect of shattered working lives as the enormity of the devastation became apparent.

Media teams encamped at various vantage points, with platform hoists to gain an aerial view of the incident.

Evacuation Centre

M1 Junction 8

Evacuation centres and hotel accommodation were quickly organised for displaced residents, many of whom were still in shock and fearful for the safety of their homes and possessions. Local stores and supermarkets responded to the sudden demand for food and clothing essentials and boarding-up materials.

Teams of council workers, volunteers and glazing companies worked throughout the day to protect and secure hundreds of properties.

Police tried to control lengthy traffic jams that built up around filling stations as panic buying began. Many of these outlets had to be closed down to eliminate bottlenecks; the police had far more urgent duties.

CLOUD OF DOOM?

The bright dawn sky provided a dramatic backdrop to the writhing column of grey-black smoke that billowed and towered above the stricken site. Clearly visible for miles, this frightening spectacle was for days on end a graphic reminder of devastation and uncertainty.

As the cloud rolled South East across London, then South West across England towards France when the wind changed, Health Chiefs issued warnings to those in its path. Asthma sufferers were advised to stay indoors, and farmers were instructed to keep livestock under cover, where possible, to prevent contamination from entering the food chain.

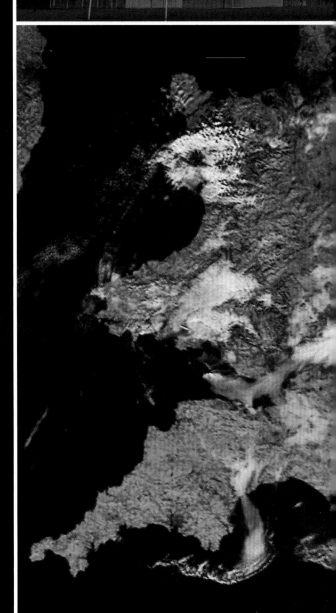

In the event, the feared ecological disaster did not materialise. The extremely-high fire temperatures caused the smoke to puncture the inversion layer (see page 127) and climb to a great height before dispersing, and there was no rainfall to bring soot particles back to earth. Slight pollution and mild respiratory irritation was the only actual consequence.

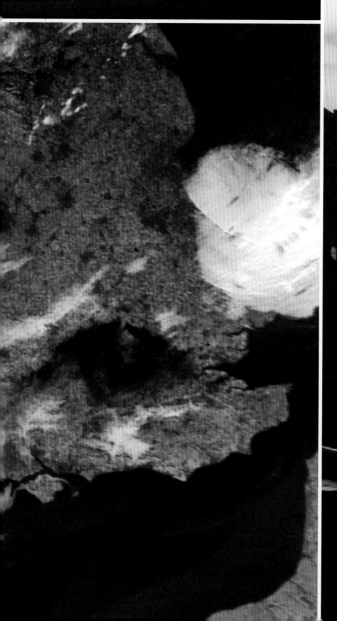

Back at Hemel Hempstead there were major decisions to be made.

A fire does not necessarily have to be put out; if safely contained it may be left to burn itself out. Here there were environmental issues either way – smoke pollution or water pollution; there were safety issues; there were political issues, including the smoke pall heading for Europe.

This fire had to be fought.

THE BATTLE BEGINS

The UK's major incident Command System – Bronze, Silver and Gold – is structured to facilitate instant response and co-operation at national level in a crisis.

BRONZE Command was set up near to Buncefield. This covered the actual fire fighting, constantly assessing risks and changing strategy where necessary, and calculating everything that needed to be brought in to fight the fire.

SILVER Command was set up at Watford Police HQ. This was in close collaboration with Bronze throughout the incident, and was responsible to locate and provide every piece of equipment and back-up that was requested. Silver was also in close touch with Gold, who had ultimate responsibility for the tactical response to the incident.

GOLD Command was set up at Welwyn Garden City Police County HQ. This consisted of the County Chiefs of Fire, Police and Ambulance Services, the Chief Executives of Health and Safety and the Environment Agency, and the Government represented by the Office of the Deputy Prime Minister. Safety, environmental and political issues can come into conflict in a major incident, and Gold provided immediate instructions in any particular situation. Gold also handled the media conferences.

"60 litres of produced foam per m² of burning substance for a minimum of ten minutes is needed to bring a fuel fire under control."

As the major incident response swung into action, equipment, firefighters and foam concentrate were called on from all over the country.

From midday on Sunday this firefighting armada began to assemble on the closed-off approach roads to the M1.

"There is no point in attempting to fight the fire too soon, until you know you've got enough foam to put it out – you can start to deal with it, but by the nature of it and the heat of it you will just get reignition if you can't cover it all."

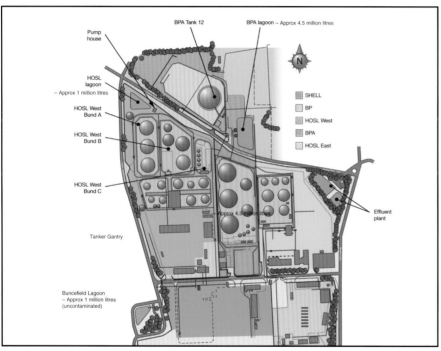

Little usable on-site water remained but an off-site lake holding 40 million litres of water from winter run-off was located just over a mile from the fire. It was 40 metres from the main road and was surrounded by fencing, a hedge and light woodland – but Bronze had to have this water and Silver made it happen.

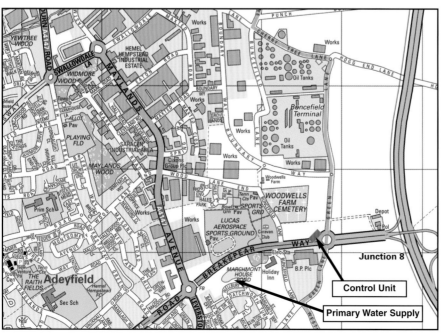

Throughout Sunday the area beside the lake became a hive of activity as men and equipment poured in – a large crane, a JCB, hardcore, high-volume pumping equipment… and a temporary roadway appeared during the night.

In the summer months this lake can be virtually dry…

One uncontaminated on-site reservoir remained. While other supplies were awaited, measures were taken to contain the fire and protect 6 intact aviation fuel tanks. From midday on Sunday a water curtain was put in place using fixed "ground monitors".

To set these up was an extremely hazardous task; firefighters worked between a wall of fire and 12-metre high tanks of warm fuel that could have exploded at any moment. As soon as the monitors were directed and operating the fire crews pulled back to safety.

Throughout Sunday afternoon and through the night the fire raged on. Contained and monitored, it was allowed to blaze but could not be fought and beaten before every element of the battle-force was ready and in place. The time would come...

Other vital preparations during Sunday and Sunday night included the construction of a temporary dam on the North East edge of the site. A major environmental concern was the large volume of waste water and foam mixed with toxic fuels and chemicals that would be produced by the firefighting. As the Buncefield site slopes towards the immediate vicinity of the underground water sources supplying North West London's drinking water, the containment of this poisonous cocktail was crucial. The plan was to pump the run-off into the secure concrete bunds that surrounded the undamaged storage tanks; the dam would act as a long-stop.

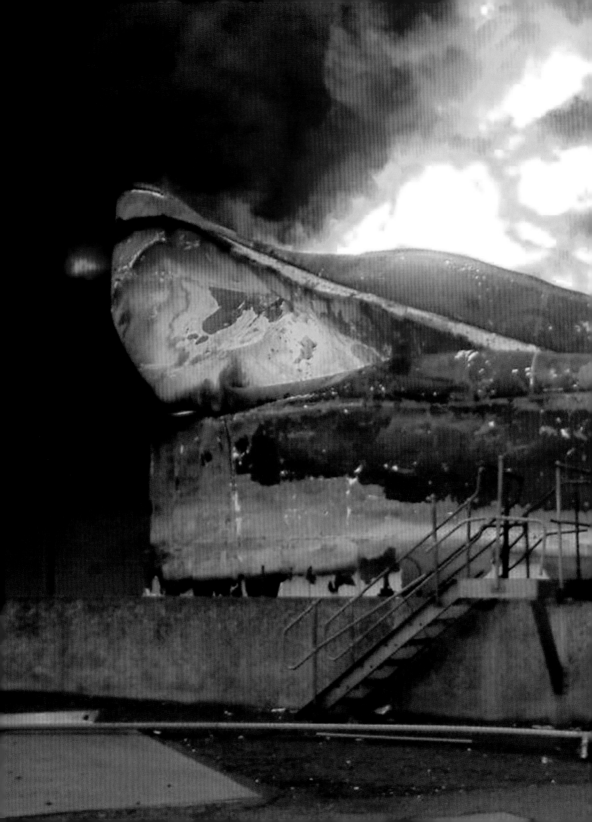

At 06:30 on Monday, with the temporary road complete, Silver brought in emergency hoses and high-volume pumps. This equipment retained by the Home Office for national emergency is designed to shift vast quantities of water. Running in parallel for over a mile, 12 hoses, each with a diameter of 9 inches, delivered 40,000,000 litres of water throughout the duration of the fight.

At 07:16 the system was ready; it had taken just 46 minutes to assemble…

PROMOTING SAFETY AT WORK **HOSL** RE OIL ITED

WARNING
Automatic Barriers

Access by Card Only
dmittance to unauthorized vehicles
ot pass until barrier is fully raised

15

Early on Monday morning, while preparations neared completion for the main attack on the North side of the site, Command decided to remove the major risk of the abandoned road tankers under the loading gantries on the South side.

From the outset these had presented a very major hazard; potentially they were flying bombs with a completely unpredictable trajectory. Still using on-site water supplies, a foam curtain was set up to cool these tankers and the fuel tanks directly behind the loading gantries to prevent their ignition. This was a containment measure; the real fight was still ahead.

Foam supplies were steadily building up, arriving in bulk tankers or lorries carrying 1000-litre containers. More was known to be on the way; factory production of the concentrate had been stepped up and supplies donated from firefighting forces from all over the country were "blue-lighted" by police escorts for hundreds of miles. Government intervention ensured a bulk shipment of foam en-route to Scandinavia was diverted back to the UK.

The scene was set – at about 8am on Monday the great fight-back began.

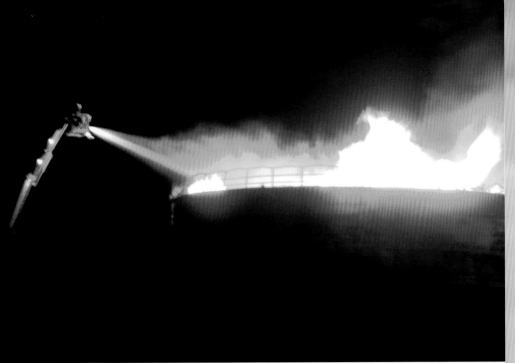

Specialist firefighting equipment came into its own, operated by highly-trained and fearless firefighters. Foam tenders sent by oil companies were especially effective because of their mobility in action around the site. The huge capacity of the foam cannons began to tell; capable of delivering 32,000 litres of mixed foam per minute, and throwing from a safe distance, they gradually checked and beat back the heat and flames for the first time.

As the onslaught continued the growing blanket of foam was steadily smothering the flames by cutting out their oxygen supply. Working in constant extreme danger, firefighters described their experiences as "exhilarating" and became increasingly reluctant to go off-duty. For them this incident had become both a deep personal involvement and a cohesive team effort that was hard to walk away from.

Early on Monday afternoon site conditions suddenly deteriorated. With firecrews working in its vicinity Tank 8 began to collapse, and blazing fuel surging over the bund wall threatened men and equipment. A rapid withdrawal from this area escalated into a full site evacuation to a safe position pending a fresh assessment.

To pull back was against every firefighter's natural inclination but a commander's duty to his men will not allow needless risk to life.

Running on adrenalin, surrounded by danger and devastation, lacking sleep and suffering the occasional set-back, the firefighters were under immense strain and at times morale dropped. The burger van – initially supplied by the Salvation Army and then by a commercial company – became an important chilling-out zone where officers could bond with their crews and take a physical and mental break from the inferno. Teams from all over the country could mingle, swap stories and regain strength before returning to the fight.

Firefighters moved back onto the offensive on Tuesday morning after new hazard assessments by experts and specialists during Monday night. Hastily-abandoned hoses and a fixed monitor had been badly damaged and some tanks had reignited. However, resuming battle with new equipment, firefighters restarted the constant bombardment of foam onto the blazing fuel. By Tuesday afternoon several tanks had been extinguished and definite progress was evident.

Superheated concrete bund walls were scarred, weakened and ripped open, leading to leakage and possible collapse. At times flaming tidal waves of fuel forced firefighters to 'vacate the immediate area.'

Acetylene gas cylinders exploded from overheating and became crazed missiles without warning. There was no way of knowing where these were and when they would explode.

The storage tanks, designed to crumple or fold inwards in an 'ordinary' fire, were liable to explode or collapse dangerously as a result of the initial blast. At times a threat of a further explosion forced firefighters to retreat to a safe distance.

The blanket of foam that successfully damped down the fire became a major hazard in itself. Every inspection cover and manhole on site had been blown out, creating hidden pitfalls everywhere.

RISK

Working in such extreme conditions became an intensely risky operation. Not having encountered such conditions before, the firefighters' greatest hazard was the inability to predict the behaviour of the fire – fighting against the unknown. There was never a safe place or a safe moment and it is nothing short of a miracle that the only injuries were very minor.

Each extinguished tank remained a high-risk hazard. As foam was blown away by the wind...

...or broke down naturally, hot fuel was exposed and reignition could occur in a matter of seconds...

...escalating into a boiling furnace again. This now threatened nearby extinguished tanks and pipelines.

The foam blanket was reapplied. A constant watch was required over the whole site to keep control of the battle.

The oily water pumped back into the bunds showed softened images of the fire that was now in its death throes. Burn-back continued to be an ever-present risk if the foam cover was disturbed, but the firefighters had the upper hand and the end was now in sight.

The front page of the local Gazette captured the initial relief of the whole community. Firefighters retired, justifiably proud but always ready for the next call.

The oppressive black cloud had at last disappeared, but the remaining uncertainty for the town was symbolised in the gaunt wreckage of a storage tank posing a stark question mark against the winter sky – what now for Hemel Hempstead?

DEVASTATION

As business owners were finally allowed to return to their properties the public was able to see the real extent of the damage for the first time. All caused by the initial explosion that shook the town, buildings within a short radius were severely damaged while those surrounded by the leaking vapour cloud were almost totally destroyed. This chaos of strewn debris graphically denoted the point where the vapour cloud may have ignited – the Fuji Film and Northgate IT car parks. These two buildings were the most severely damaged.

About 120 staff worked at the Fuji film premises. One security guard was on duty and recalled a very strong smell of fuel just moments before the blast hit the building. He was shocked and disorientated but only slightly hurt.

Staff from Golden West Foods, across the road, used the Fuji car park at weekends. Their 6 am shift had just arrived – for some reason no-one was late that morning, and by 06:01 every car, just vacated, was totally wrecked.

On Monday Fuji negotiated alternative leases on premises in Rickmansworth and Milton Keynes and by Wednesday their staff were installed.

Northgate Information Solutions employed about 400 here. Four were on site, miraculously suffering only cuts and bruises from flying debris. This was the only building outside Buncefield to catch fire.

Fire crews searching this building reported utter destruction – collapsed concrete floors and ceilings, ravaged interiors, and every window shattered with shards of glass embedded in walls and pinboards.

Northgate suffered the greatest dislocation of staff, but within days the majority were accommodated elsewhere. The company already had a carefully-planned and well-rehearsed business recovery strategy which ensured that disruption to clients was minimal. Northgate I.T. has since become a benchmark for industry contingency planning.

To a greater or lesser extent nearby local companies suffered severe or total disruption – loss of premises, loss of equipment, loss of stock, loss of trading continuity… but, unbelievably, no loss of life. Some buildings were completely laid waste by the blast, some had fractured exteriors and interiors and some appeared safe but had suffered enormous vibration. Over 370 businesses and over 3,500 workers were affected.

For days many were not even permitted to enter their buildings to establish what might be retrievable; owners did not know whether their business still existed and employees did not know if they were jobless. For thousands life would never be the same again – the familiar work environment, its staff, its daily routine, seemingly so stable but suddenly so vulnerable, and so nearly a scene of unimaginable horror if this had ocurred during a working day.

Ironically Buncefield Depot itself employed very few; it was beyond its fences that the commercial impact was so cataclysmic.

For those living near the site, home was no longer a place of safety. The blast wave that ripped across the town had blown in windows, doors and patio doors, cracked walls, damaged roofs and brought down ceilings and fixtures. Flying glass had ripped furniture, clothes and possessions. Houses were vulnerable to the weather and even to looting. Those not evacuated began the long process of recovering from shock, assessing the situation and starting the lengthy clear-up and patch-up operation.

Sixty members of the public received hospital treatment for minor injuries.

Local shop owners were disrupted by broken front windows. Shops generally suffered loss of trade in the normally busy period in the run-up to Christmas.

Fred and Elizabeth Mills lived in the small cottage less than fifty yards from tank 12 – the biggest tank on site. Their cottage was practically destroyed by the blast. Fred escaped possible death by moments – he got up to make a cup of tea and seconds later bricks and debris rained down on his bed.

Ian Silverstein lived in 'High Grange', a large house approximately four hundred yards away from the site. Ian spent years renovating his house into a luxury dwelling on the edge of the countryside, but also on the edge of a fuel depot! The early morning explosion transformed his home into a wreck and narrowly missed seriously injuring or killing its inhabitants as they slept.

High Grange

The Mills' Cottage

Fuji

110

Wickes Storage Depot

New Distribution Unit

Northgate IT

Triumphant Heroes

Risking their own lives to save others, firefighters from all over the country joined with local forces, walking into a terrifying towering inferno whilst the rest of the town ran for their lives. Retiring fifty-nine hours later, triumphant at having extinguished what the country called "the biggest fire in peacetime Europe" – the biggest fire they will probably ever see – this team of heroes are justifiably proud. An outstanding success, this feat could not have been achieved without the loan of specialist equipment, expertise and sheer manpower from other firefighting units.

To everyone of you and all the other services involved I send me deepest admiration for the courage and skills and sheer guts it takes to tackle such a situation. You can all be proud of yourselves.

Dear Firemen, thank you for putting out the serious fire in Buncefield. I heard the loud bang and looked out of my room window, when I saw the sky I was a bit scared. I think you are all very clever and you have superb ideas to do things like that fire, love Edward

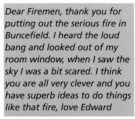

Dear Hemel Hempstead Firemen, thank you for putting out the fire at the Buncefield Oil Depot. How did you put it out? It was so big. It made me jump. Yours sincerely Kyle, aged 6.

You are a great lot – you have saved life and taken care of your men. We take your service for granted – but it is this sort of thing you have faced that wakes us up to what you face. (Brian, Essex)

We wish to send our grateful thanks to all the firemen who worked so hard to put out the fire at Buncefield. The obvious careful planning, organisation and teamwork showed us how wonderful our fire service is. A magnificent and brave effort.

Fighting around the clock on extended shifts, Hertfordshire Fire and Rescue Service also attended a total of 348 other calls throughout the course of the fire, resulting, as described by one fire chief, in "an exercise in sleep deprivation." With the fire out and the incident closed, firefighters returned to their normal shifts. Almost every day, a firefighting force, through their courage and specialist knowledge, will save lives and property, and avert serious injury. So often a firefighter's work goes unnoticed – this time the whole world saw. This was their moment, their glory.

Dear Hemel Hempstead fire fighters, thank you for putting out the fire at Buncefield Oil Depot. Thank you for saving my Aunty because we were worried about her. Yours sincerely, Lewis.

I have written to the Prime Minister to say that every fire fighter should be honoured for bravery in the oil depot fire. All of you went beyond the call of duty and as human beings all of you are second to none in this country. (local resident)

I am sure that bickering, buck-passing and incriminations will boil up to generate as much heat as the fire at Buncefield, However, I want to put on record my unstinting praise and admiration for your men, who beat the raging inferno in such a heroic manner.

To the fire service of Hemel Hempstead, every day you put your lives at risk, Buncefield is just one example. Thank you for keeping us safe. (Boxmoor resident)

Thank you all for the fantastic work you have all done to combat the awful fires at Buncefield. I know you've trained to fight fires but no training could have prepared you for Sunday. It must have taken great courage to combat the fires, and to put them out in just 3 days is nothing short of miraculous.

Blue Watch, Hemel Hempstead Fire Station were first on site. These initially faced the awesome power of the fire and the intensely risky search for survivors.

Forces from all over the country joined local crews with the same determination to get the job done...

Hertfordshire
London
Bedfordshire and Luton
Essex County
Norfolk
Suffolk
Warwickshire
West Midlands
Cumbria
Greater Manchester
North Yorkshire
Humberside
Buckinghamshire
Hampshire
Kent
Oxfordshire
Royal Berkshire
Surrey
Derbyshire
Northamptonshire
Nottinghamshire
Mid and West Wales
South Wales Fire Service College.

God's Mercy

"yes it was God's mercy... I think a lot prayed that day that perhaps wouldn't have done otherwise..."

"I do not regard the explosion as a catastrophe. I think it was a miracle – no one was killed..."

"you've probably been told this but I think it's a miracle that nobody died. I think if it had been 9 o'clock on a Monday rather than a Sunday, it would have been a different story... it was as if God's hand was over it..."

"so thankful for the Lord's mercy especially in the timing of it, as there was no loss of life at all and very few injuries..."

"from the time of day it happened there's nobody can say there isn't a God in heaven..."

...and the cause?

The HSE's investigation into the incident indicates that the cause of the explosion was the overflow of Tank 912 in Bund A on the HOSL West site. Standing 18.2 metres high with a diameter of 25 metres, this tank had a capacity of over 6,000,000 litres.

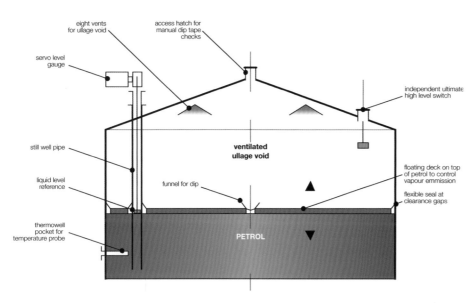

At the time of the incident fuel was being delivered to Buncefield through three pipelines. One of these, the T/K South line from Lindsay Oil Refinery on Humberside, began delivering a batch of 8400m³ unleaded petrol at around 19:00 hours on Saturday. This delivery was split between Buncefield and BPA's site at Kingsbury, near Birmingham, giving a flow rate to Tank 912 of 550m³/hour.

Records from Tank 912's ATG (automatic tank gauging) system suggest an anomaly or malfunction. From 03:00 hours on Sunday the system indicated that the volume remained static at about ²/₃ full, below the level at which the ATG system would trigger alarms. However the records also indicated that the inlet valve to Tank 912 was open at the time of the incident, suggesting that the tank was still filling.

Approximately 7 minutes before the incident the Kingsbury line was closed, leading to an increase in the flow rate to Tank 912 of 890m³/hour.

At around 05:20 hours on Sunday petroleum began to overflow through the airvents at the top of the tank. Fuel flowed down the sloping roof splashing over the deflector plate on the rim of the tank (this plate is designed to direct cooling water down the outside of a tank in the event of a fire). Flowing down the side of the tank, fuel then hit a 'wind girder' spanning the circumference of the tank about ¹/₃ from the top, causing further splashing and cascading to the ground.

This agitation of the leaking fuel created a volatile fuel/air vapour which, according to CCTV records, by 05:38 hours had begun to seep from the North West of bund A at a height of about 1 metre. The fuel, pumped in at a temperature of about 7 - 8°C, had condensated in the freezing air temperature creating a visible mist.

By 05:46 hours, the vapour was 2 metres high, and by 05:50 hours was flowing off the site near the junction of Cherry Tree Lane and Buncefield Lane, and into the Northgate and Fuji carparks.

By 06:01 hours, a total of 300 tonnes of unleaded fuel had poured from the top of the tank into its bund. This hydrocarbon-rich vapour cloud, which now covered an area of 80,000m², ignited with resultant overpressures in the order of 700 - 1000mbar, a destructive force 20 times greater than any other previously-experienced vapour cloud explosion.

The actual ignition source has not been established with certainty, but could have been the HOSL West pumphouse, the Northgate emergency generator cabin or the running engines of nearby cars.

FURTHER INFORMATION

- Oil – from underground to petrol tank
- Buncefield – a planning dilemma
- A depot in action
- The explosion and fire – a meteorological phenomenon
- 1985 – a near miss

OIL

Oil is essential to modern life; it provides the fuel for most transport and the raw materials for a multitude of other products, chemicals and plastics.

Crude oil is a fossil fuel, formed by the buried remains of billions of marine animals and plants. These sank to the sea bed when they died, gradually accumulated and were then buried by heavy sediments. The resultant pressure and heat changed the organic matter into the hydrocarbons we know as gas and oil. Subsequent movements of the earth's crust created traps where the oil and gas collected.

Geological surveys are used to detect the likely existence of oil and gas. The most valuable are seismic surveys which use sound to produce shock waves that pass into the earth's crust and are reflected back from the various rock layers.

Only drilling can prove the actual existence and extent of reserves. Once located, extracted oil is piped from the production platform to an onshore terminal, then transferred to sea tankers for delivery to refineries around the world.

At the refinery, different crude oils are blended and then passed through the stages of distillation, catalytic cracking and other special processes to maximise the proportion of actual petroleum obtained. Testing procedures are carried out at the refinery to ensure the quality of each product.

Depots such as Buncefield are essential. The supply of crude oil is subject to many unpredictable geographical and political factors, and the provision of storage facilities enables the industry to control or minimise the effects of fluctuations in the supply-chain.

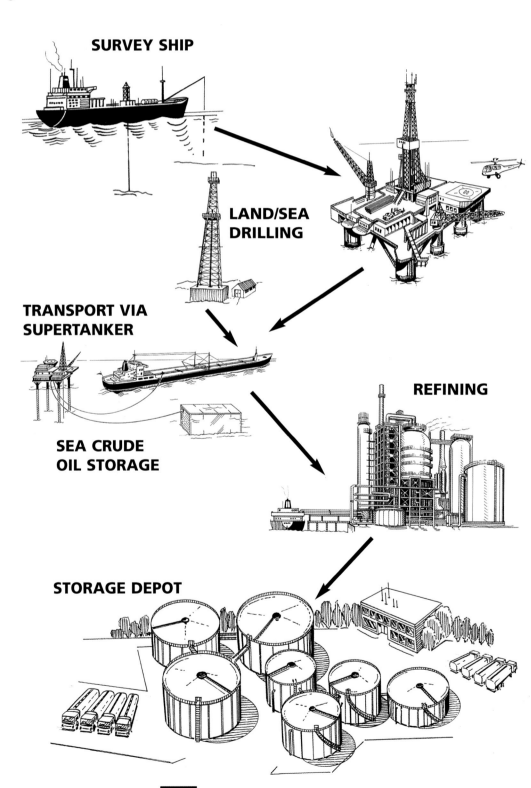

SURVEY SHIP

LAND/SEA DRILLING

TRANSPORT VIA SUPERTANKER

SEA CRUDE OIL STORAGE

REFINING

STORAGE DEPOT

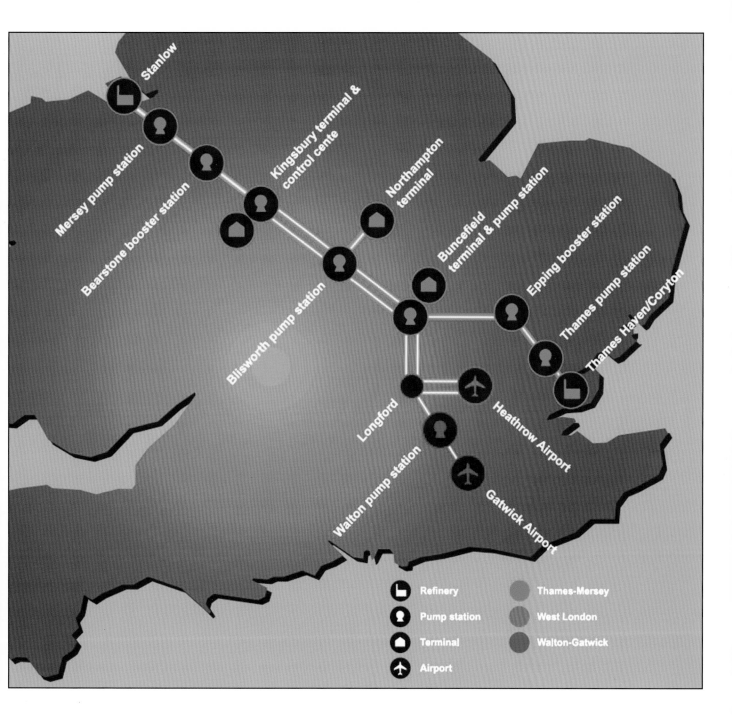

Map legend:

Refinery		Thames-Mersey	
Pump station		West London	
Terminal		Walton-Gatwick	
Airport			

Map labels: Stanlow, Mersey pump station, Bearstone booster station, Kingsbury terminal & control cente, Blisworth pump station, Northampton terminal, Buncefield terminal & pump station, Epping booster station, Thames pump station, Thames Haven/Coryton, Longford, Heathrow Airport, Walton pump station, Gatwick Airport

Petroleum products are moved from source to destination via a national pipeline network. Stretching from the Shetland Isles to Dorset, this vast structure efficiently carries fuel from oil fields to refinery to depot or terminal.

One pipeline will carry many types of fuel. Inevitably, as one batch of product follows another, mixture or interface occurs. This is managed by pumping the first one hundred gallons of each batch into a 'slop' tank. When full, this mixture gets sent back up the pipe network to a refinery for refining. Pipework is maintained and kept clear by pumping steel balls along with oil products.

Marker posts up and down the country indicate the path of a pipeline. These posts can be monitored by air, and any digging works in the vicinity of the network can be identified and controlled.

As can be seen from the map, Buncefield – the fifth largest oil depot in the UK – plays a central part in the distribution of fuel around London and the South East.

A Planning Dilemma

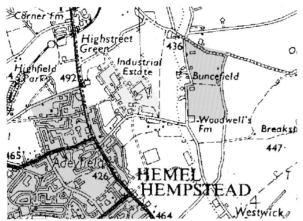
Buncefield 1958

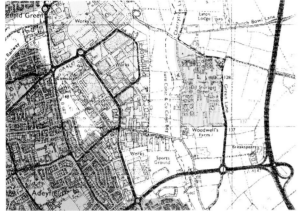
Buncefield 1980

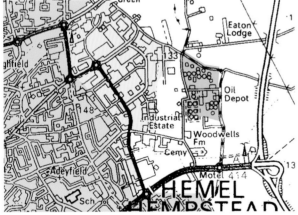
Buncefield 1988

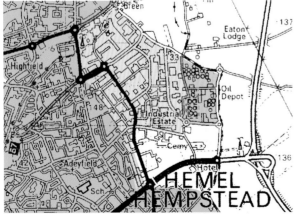
Buncefield 2001

All maps above reproduced by kind permission of Ordnance Survey © Crown Copyright. All Rights reserved. Licence No. 100045869

Hemel Hempstead became one of several New Towns that were designated for rapid residential and business development after the war, to relieve overcrowding in London. An important feature of its planning policy was the establishment of a separate industrial estate on the North East boundary of the town, providing a definite demarcation from residential areas and close proximity to the M1 motorway.

Clearly this segregated and accessible location appealed to the oil industry as ideal for the establishment of a storage and distribution depot, well-suited to both the pipeline supply network and the road-tanker delivery operations.

The initial development of the depot was the subject of normal planning procedures in the mid-1960's; the Ministry of Housing and Local Government granted consent after a public enquiry and the first part of the Buncefield Depot was opened in July 1968, primarily to provide aviation fuel for Heathrow Airport.

The site covered 74 acres, and was relatively isolated; a band of open land, 300-400 metres wide, remained between the depot and the then eastern limit of the industrial estate.

Subsequently, with land in short supply and values soaring, the local Planning Authority has been under relentless pressure to release surrounding areas for development. Thus we have today's predicament: Buncefield has expanded to its present capacity of 29 storage tanks covering over 100 acres, the industrial estate has advanced right up to the Buncefield fences and modern residential areas have crept uncomfortably near to both.

Coloured maps demonstrate the change in residential and industrial growth around the site over the years.

"Not In My Back Yard" - but whose back yard?!

Buncefield Oil Depot 1996

Buncefield Oil Depot December 19th 2005

Buncefield
A Depot in Action

Buncefield terminal receives, stores and distributes finished petroleum products - Aviation Fuel, Diesel, Petrol, Kerosene and Gas Oil. In addition to the storage tanks and loading gantries the site facilities include administration offices, control rooms and laboratories.

Aviation fuel is sent by pipeline to storage facilities at the major airports. Other products are distributed by road, necessitating about 400 tanker collections every 24 hours, round the clock.

All these operations are protected by rigorous health and safety procedures, constantly reviewed and rehearsed, with extensive fire-fighting instruction and training. The companies based at Buncefield work closely with the local Fire Brigade, the Health and Safety Executive and the Environment Agency.

Filling-up procedure – In the words of a tanker driver

"Well there's three drivers for every tanker and they keep them on the road 24/7. The first thing you do is check the vehicle: that's tyres, lights etc...You go to the office and check the vehicle, then make sure you've got the right PPE, you check it off, then what happens is that you load up. But before you load up with fuel you have to connect the gantry to the scully: it's an earthing system because otherwise you could get a static shock that could ignite it all.

So your brakes are on and the scully is fitted and you can then fit the Vapour Recovery system. This is a four inch hose because the fuels are being pumped so fast vapour is being blown out that can't be released into the atmosphere because of pollution, so the vapour goes into the system and around, not into the atmosphere: it goes into the hose connected to the lorry and then you're safe to load.

There are 6 loading arms and you can do 3 arms at once, taking about 5 minutes and then the next 3; and the whole 38,000 litres takes about 15 minutes to load - that means you are on and off the gantry in 15 minutes.

To load, you go to the computer and punch in who you are and what vehicle and it will then tell you what to take.

I've been on this for 30 years now. Of course, I had 3 months training - that was petrol, diesel, gas, bitumen - the lot."

What was in the tanks? The site can store a potential 177 million litres of fuel, but not all tanks were full at the time of the incident. The magnificent efforts of the fire fighters also prevented the spread of fire across the site; many millions of litres were saved from combustion. Details of the tanks on site are given opposite.

Tank number	Max gross volume (litre)	Diameter (m)	Height (m)
301	3,505,634.00	21.30	13.20
302	3,505,511.00	21.30	13.20
303	5,781,125.00	24.40	13.30
304	4,587,171.00	24.40	13.30
305	1,679,385.00	14.60	11.40
306	1,683,693.00	14.60	11.30
307	2,012,327.00	14.60	13.20
901	2,183,209.00	17.10	13.20
902	2,917,344.00	19.50	13.20
903	2,921,539.00	19.50	13.20
904	1,636,081.00	14.60	13.20
905	2,912,114.00	19.50	13.20
906	1,999,737.00	12.80	13.20
907	2,720,309.00	12.80	13.20
908	1,187,806.00	10.90	11.40
909	1,712,812.00	10.90	11.40
910	6,050,115.00	25.00	18.20
911	6,048,563.00	25.00	18.20
912	6,049,173.00	25.00	18.20
913	3,373,341.00	25.00	14.10
914	6,055,535.00	25.00	18.20
915	6,044,255.00	25.00	18.20
916	1,480,494.00	12.50	15.10
4	12,908,706.00	36.50	12.80
5	12,906,192.00	36.50	12.80
6	12,905,499.00	36.50	12.80
7	5,434,192.00	24.00	12.00
8	545,222.00	24.00	12.00
9	389,982.00	7.60	9.20
10	390,164.00	7.60	9.20
11	389,722.00	7.60	9.20
12	18,998,589.00	44.00	12.80

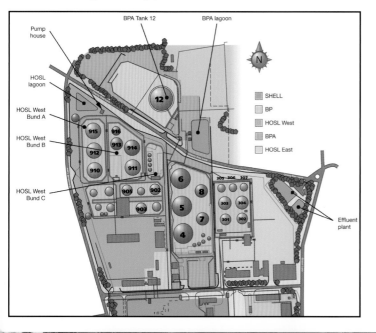

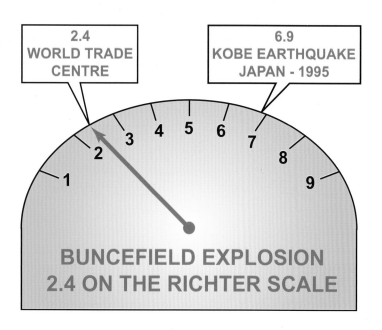

2.4
WORLD TRADE
CENTRE

6.9
KOBE EARTHQUAKE
JAPAN - 1995

BUNCEFIELD EXPLOSION
2.4 ON THE RICHTER SCALE

The Buncefield blast measured 2.4 on the Richter scale. This is equivalent to a small earthquake. It is interesting to note that the collapse of the World Trade Centre, 11 September 2001 measured 2.4 on the same scale.

Travelling at the speed of sound (approx 743 mph) the blast wave spread across the UK. Heard by thousands – a fact which put many people in touch with the disaster – the blast destroyed whole offices in Hemel Hempstead, shook buildings 25 miles away, and rattled letter boxes on the east coast.

A meteorological phenomenon called an inversion provided a tunnel effect, channelling the sound at ground level for over 100 miles. Those so far away would have heard the blast up to 10 minutes after the initial explosion. Shock waves can travel up to 30 times faster than sound, resulting in some feeling the explosion before it was heard.

Inversion

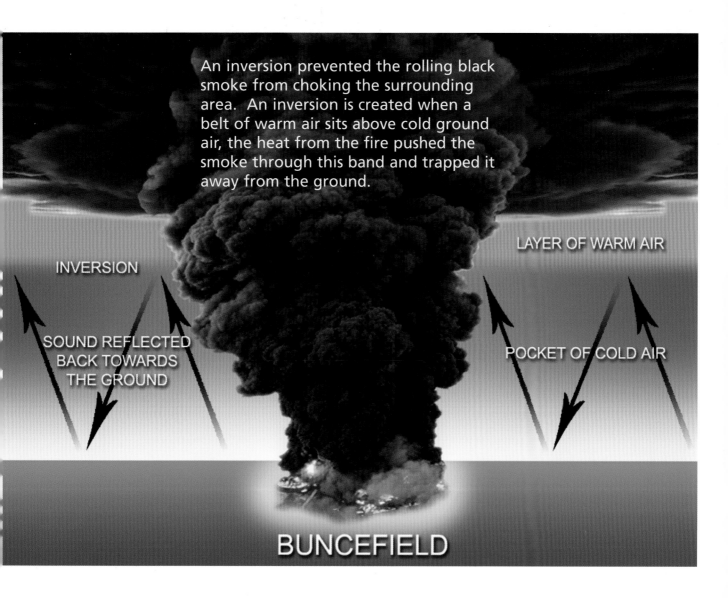

An inversion prevented the rolling black smoke from choking the surrounding area. An inversion is created when a belt of warm air sits above cold ground air, the heat from the fire pushed the smoke through this band and trapped it away from the ground.

LAYER OF WARM AIR

INVERSION

SOUND REFLECTED BACK TOWARDS THE GROUND

POCKET OF COLD AIR

BUNCEFIELD

The inversion contributed to the relatively minor pollution from such a vast oil fire. The dispersal of smoke at a high altitude prevented a widespread environmental distaster. Photographs taken of the smoke demonstrate this effect taking place.

By comparison it is interesting to note the smoke behaviour from the Northgate fire; pages 38/39 and 46/47 show this cooler smoke drifting and remaining closer to the ground.

Just a few yards— from major disaster

mid 1985

OIL GIANTS Mobil were fined thousands of pounds on Monday for breaching health and safety regulations after a fire at their Buncefield depot which could have developed into a major catastrophe.

Tanks holding millions of gallons of petrol could have blown up, but the potentially massive disaster was averted by firemen who managed to put out the blaze and prevent it from spreading.

Firemen had to carry out more work to make the area safe when further tests revealed the concentration of petrol vapour from a leaking pipe was so high that there was still the danger of a huge explosion.

The frightening incident happened on the Mobil site on June 8 this year when the fire started near tanks after workmen had been carrying out modification work on pipes with welding gear.

A major alert was raised when Mobil staff couldn't bring the fire under control and fire engines from Hemel Hempstead and Garston raced to the scene and managed to put it out within 13 minutes.

Magistrates in Hemel Hempstead heard that Mobil had started the work without the permission of the brigade and had failed to take enough precautions to prevent a fire or explosion.

Sub Officer Alan Lodge told the court that when firemen arrived they discovered the fire inside a bund (fire break) wall adjacent to tanks. The blaze had been extinguished with a foam blanket. He said he had been told the fire had been caused by a leaking kerosene pipe and it was only afterwards that he discovered it was a petrol pipe. He added he probably wouldn't have sent his men in if he had known.

Assistant Divisional Officer Roland Ginger said he issued a warning after the fire had been put out when an explosion-meter reading revealed an equally dangerous situation with a large amount of petrol vapour in the area and a "potentially explosive and hazardous situation."

He said he discovered petrol was still leaking

● The Buncefield depot on the day of the fire when a major disaster was narrowly averted.

Oil company fined £5,500 for breaking safety regulations

from the pipe and that hundreds of gallons had leaked on to the floor. He had been amazed, he said, that people had been working on it when it should have been dry and vapour free.

He said he had also been alarmed to see an open lid on an adjacent empty tank which was "potentially lethal".

If the vapour had ignited the tank could have exploded possibly rupturing other tanks.

Deputy Senior Fire Protection Officer Ivan Reader said if they had known about the proposed work he would have suggested it was done outside the bund or under constant supervision, with meter readings, a foam blanket laid down and with fire equipment immediately available in case of a fire.

He said that from his investigations there were a number of possible causes of the fire. The most likely was an arc from two leads used for welding which had been left inside the bund on damp ground.

If the fire had spread the ten yards to the empty tank the "consequences could have reached major disaster proportions" and would have taken hours if not days to put out, he added.

Mr Roger Santon from the Health and Safety Executive said no hot work should have been carried out inside the bund unless it was "impossible" not to and added: "The incident could have certainly developed into a major catastrophe."

Mobil project engineer Barry Collison said it would have been "impractical" to move the pipe outside to do the work. The pipe was 59 feet long and weighed two tons and the hot work involved was "relatively small and quick".

He said he had told the contractor to have fire extinguishers nearby to dampen the floor down with water and to use an "elephant" cover when doing the work. Despite investigations they could not find why the pipe had leaked.

Counsel for the company, Mr Richard Rundell, told the magistrates the company had taken all reasonable precautions.

Mobil, who admitted failing to give notice of the work, were found guilty of failing to maintain a valve on Tank 4, failing to take all due precautions to prevent accidents by fire and explosion and allowing the use of welding and burning equipment in an area which had not been made safe and certified free of petrol vapours.

They were fined a total of £5,500 and ordered to pay the prosecution costs of £500.

A newspaper report, unearthed after the disaster, uses uncannily familiar language in its description of a previous blaze which was swiftly brought under control.